Francisco Thiago de Albuquerque Aragão
Diego de V. Lourenço
Thales V. de Araújo Viana

Chlorophyll content at the end of the caupi bean cycle in partnership with

Francisco Thiago de Albuquerque Aragão
Diego de V. Lourenço
Thales V. de Araújo Viana

Chlorophyll content at the end of the caupi bean cycle in partnership with

with panicum maximum grass, under planting blades and times

ScienciaScripts

Imprint

Cover image: www.ingimage.com

This book is a translation from the original published under ISBN 978-3-330-75804-9.

Publisher:
Sciencia Scripts
is a trademark of
Dodo Books Indian Ocean Ltd. and OmniScriptum S.R.L publishing group

120 High Road, East Finchley, London, N2 9ED, United Kingdom
Str. Armeneasca 28/1, office 1, Chisinau MD-2012, Republic of Moldova, Europe
Printed at: see last page
ISBN: 978-620-8-29531-8

SUMMARY

To my parents, my main reason for being here, for supporting me, for encouraging me, for all the support given and for the advice, consolation, conversations and encouragement throughout this journey.

THANKS

To God, for always being by my side, not abandoning me when I need it most, source of love and wisdom, my protection.

To my mother, Maria Santissima, for always protecting me from all dangers and guiding me along the right path.

To CNPq, for the financial support with the maintenance of the scholarship.

To the Federal University of Ceará, for the opportunity and the infrastructure to carry out this course.

To my parents, Chagas and Cristina, the reason I'm here, for all the love and dedication they've always had for me.

To Maria Edirane, for all her maternal love for me, for raising me and giving me so much love.

To my grandparents Maria do Socorro (In Memoriam), Raimundo Alberto, Maria Neuma and Manoel for all the love and affection they have shown me.

To all my family members, especially my uncle Jorgelanio (In Memoriam), for all the support he gave me while I was here on earth.

To my aunt, Maria da Conceiçao, who, despite the distance, is present at every moment of my life.

To Prof. Dr. Thales Vinicius de Araùjo Viana, for his excellent guidance and example as a professional.

To Prof. Dr. Francisco Marcus Lima Bezerra, for accepting to be a member of my panel and for all the knowledge he passed on to me during my undergraduate studies.

To Msc. Chrislene Nojosa and Régis Santos for all their support, co-supervision, advice and for making the writing of this work possible.

Jùlia Medeiros, italo Oliveira and Robevânia Borges for their time, valuable contributions and suggestions.

My friends Keivia Lino and Patricia Alcântara, for all their help during the execution of this work.

To my great friend Tiago Cavalcante, for his valuable contribution during the experimental phases of this work, for all his help and dedication.

To Agronomist Lais Monique, for all her help, for recommending and introducing me to my advisor.

To the scientific initiation scholarship students, Samuel de Sousa and Maiara Pereira, for being present during the conduct of the experiment, helping whenever necessary.

To my friends Danyelle Mauta, Amanda Pinheiro, Joyce Cristine, Camila Castro, Daniele Varelo, Fernanda Helena, Ingrid Pinheiro, Cecilia Barreto, Renata Araùjo, Vinicius Bittencourt, Adriana Celentano, Clice Mendonça, Marcelo Silva, Tiago Rodrigues, Elaine Silva and Iorrana Gomes, who were part of my story and were always with me during my degree, sharing moments of joy, sadness and victory.

To the Seed Analysis Laboratory of the state of Cearà, especially to my friends Gina Karolli and Alexandre Nunes, for all the knowledge they passed on to me during my time as an intern.

To the staff of the agrometeorological station, Dona Marilac, Weverton and Ricardo for their readiness and help whenever necessary.

To Professors Patrik Pastori, Ricardo Romero, Thiago Osório, Carmem Dolores, José Carlos, Patricia Lima and Rosimeyre Melo, for having marked my degree with their love for the profession.

To the family I formed here in Fortaleza, Carmem, Natàlia and Gabriella. Thank you for always being there for me.

To Conceiçao de Maria, Kayline Macêdo and Sabrynna Britto, who, although far away, always find a way to show that they are there.

To my best friend Diego de Vasconcelos, for these seven years of friendship, for his patience, advice, help and promptness.

To all those who contributed in some way to this work, my academic victory.

"The Lord God placed man in the garden of Eden to care for it and cultivate it"

Genesis 2:15

SUMMARY

The chlorophyll index present in caupi beans in the final phase of production was evaluated in an experiment carried out in the experimental area belonging to the Department of Agricultural Engineering of the UFC, in Fortaleza - Ce, Brazil, with an experimental design in randomized blocks and subdivided plots, the plots being composed of the treatment of different blades in relation to the evaporation of the class A tank (ECA) (0.25; 0.50; 0.75; 1.00; 1.25 times the ECA) and the sub-plots were made up of grass transplanting times (0; 5; 10; 20 DAEF - Days after bean emergence.), each treatment consisting of three replications.

Observing the effect of the blades evaluated, those equivalent to 0.75 and 1.00 times the ACE stood out, which showed the highest chlorophyll index in the leaves of the Caupi bean. As for the treatment of grass insertion time, the 10 and 20 DAEF times stood out, showing that the legume's competition with the grass directly affects chlorophyll production in the bean plant.

Keywords: Crop-livestock; Planting times; Water availability;

1 - INTRODUCTION

Nowadays, there are several challenges facing agriculture, including the production of food in high quantity and quality. In this sense, the most appropriate alternative is to use production systems that intensively use the resources available in agrosystems and simultaneously improve soil quality (BALBINOT JUNIOR et al., 2009). Crop-livestock integration (CLI) can be included as one of these production systems, as it consists of diversification, rotation, consortium and/or succession of agricultural and livestock activities within the rural property, in a harmonious way, constituting the same system, in such a way that there are benefits for both, increasing productivity and reducing costs (ALVARENGA; NOCE, 2005).

In subsistence agriculture, the main products grown on farms include corn, rice, beans and cassava. Among these, in the semi-arid region, the cowpea *(Vigna unguiculata* L. Walp) stands out because it is a grain rich in protein, fiber and minerals and also because it has characteristics that are resistant to the conditions of the region and because it produces early when compared to other grains. Because of these factors, the crop is widespread in the Northeast, being grown both in rainfed conditions and in irrigated perimeters.

Agriculture depends on two essential factors of production: water and soil. The northeastern semi-arid region is characterized by intense water restrictions, which impose limits and conditions on agricultural production (BUAINAIN; GARCIA, 2015). The use of irrigation is essential to make production viable in the semi-arid northeast. However, with the current scenario of droughts that the region has been experiencing, it is essential to optimize the use of irrigation, offering only the amount of water needed, so that it is not wasted. To do this, it is necessary to know a water table that satisfies the water needs of the crops, especially in the case of intercropping.

In order to optimize space with the consortium between the two species, and knowing that the growth of the BRS Zuri grass (Panicum maximum) is faster than that of the BRS Marataoa bean *(Vigna unguiculata L),* the main objective of this work was to identify the ideal time for transplanting the grass and the appropriate irrigation level to obtain good production, with the aim of analyzing which treatment produced the best chlorophyll indices.

2 - LITERATURE REVIEW

2.1 - Crop-livestock integration

Nowadays, there are several challenges facing agriculture, including the production of food in high quantity and quality. In this sense, the most appropriate alternative is to use production systems that intensively use the resources available in agrosystems and simultaneously improve soil quality (BALBINOT JUNIOR et al., 2009).

For low-income farmers with small areas to cultivate, greater attention must be paid to the cost of production and the best use of the land. In this context, crop consortiums can become a very important practice for subsistence farming (RAPOSO et *al.,* 1997). Crop-livestock integration (CLI) can be included as one of these production systems, as it consists of diversification, rotation, consortium and/or succession of agricultural and livestock activities within the rural property, in a harmonious way, forming the same system, in such a way that there are benefits for both, increasing productivity and reducing costs (ALVARENGA; NOCE, 2005).

According to Mohamed Sallen and Fisher (1993), the alternation between activities is the main reason for the increase in productivity in the area. McKenzie et al. (1999), explain that as a benefit for crops, pasture offers improved soil fertility, better weed control, breaking the cycles of diseases and pests and for livestock, there will be an increase in the availability of good quality food for livestock, either during grazing or for making hay and silage.

Crop-livestock rotation appears to be a promising strategy for developing production systems that are less intensive in their use of inputs and more sustainable over time (ASSMANN et *al.,* 2004). According to Alvarenga and Noce (2005), the main objectives of ILP are to recover or reform degraded pastures, improve the physical and biological conditions of the soil with the pasture in the crop area, produce pasture, conserved fodder and grain in the dry season, reduce dependence on external inputs and reduce costs in both agricultural and livestock activities.

According to Andrade et al. (2003), a fundamental issue for the success of the system is the right choice of species. In the case of forage species, it is not enough for them to be shade-tolerant; it is necessary to choose species with good productive capacity, adapted to management and suited to the soil and climate conditions of the region where they will be planted. The association of legumes with the system is also important to increase the availability of nitrogen to the ecosystem.

Myers and Robbins (1991) found a reduction in the availability of nitrogen in the soil in pasture systems not intercropped with legumes, when compared to intercropped systems. Similar results were found by Binkley and Giardina (1997), but in areas of forest planted with Eucalyptus, thus demonstrating that in consortium systems, the use of legumes is a factor that contributes to successful production.

According to Bezerra et al. (2008), the main advantage of associated crops over isolated crops is that they make better use of the available resources of water, nutrients and light. Grass and legume associations in particular are complementary in their use of production factors (RESENDE et al., 1997).

2.2 - Growing cowpeas *(Vigna unguiculata* L.)

The cowpea crop is of African origin and was introduced to Brazil in the second half of the 16th century by the Portuguese colonizers in the state of Bahia and then spread throughout the country (FREIRE FILHO, 1988).

Its domestication occurred in agricultural systems composed of sorghum and millet, which predominate in the semi-arid regions of West Africa and are characterized by having one of the largest areas of cowpea production (SINGH et al., 2002). Due to its wide geographical distribution, over time it has acquired several popular names in Brazil, including feijao-macassa and feijao-de-corda, in the Northeast; feijao-de-praia, feijao-da-colônia and feijao-de-estrada, in the North; feijào-miùdo, in the South (FREIRE FILHO; CARDOSO; ARAÙJO, 1983).

The cowpea is an excellent source of protein (23-25% on average) and also all the essential amino acids, carbohydrates (62% on average), vitamins and minerals, as well as having a large amount of dietary fiber, low fat content (2% oil content on average) and no cholesterol (ANDRADE JUNIOR et al. 2002). Due to its high nutritional value, hardiness and low water demand, the cowpea is a staple food for low-income families in northeastern Brazil, and is also one of the main grains grown by families practicing subsistence farming (RIBEIRO et *al.,* 2002).

Worldwide production averages 12.5 million hectares, with Nigeria, Niger and Brazil being the main producers (QUIN, 1997). In Brazil, most of the total production comes from the semi-arid sertao of the Northeast and small areas of the Amazon (MAIA, 1996). The largest national producers are the states of Cearà, with an average production of 159,471 tons per year^{-1} , followed by Piaui with an average of 58,786 tons per year, Bahia with 50,249 tons per year^{-1} and Maranhao with 35,213 tons per year^{-1} , which also have the largest planted areas (Levantamento Sistemàtico da Produçao Agricola - IBGE, 1993-2015).

2.2.1 Weather

Studies on the physiology of caupi in Brazil are somewhat limited, with most of the information used coming from work carried out in other countries. Among the known elements of climate, the ones that show the greatest influence on the development of the crop are rainfall and air temperature, which, through climate risk zoning, make it possible to verify the viability and the appropriate time for planting the cowpea crop. Other elements of the climate that influence the growth and development of this crop are: photoperiod, wind and solar radiation (RIBEIRO, 2002).

As far as temperature is concerned, the crop adapts easily to different regions, from latitude 40° N to 30° S, adapting well to both high and lowlands in West Africa, Asia, Latin America and North America (CARDOSO et al., 2005). Bastos (1999) found that the good development of the crop occurs in the temperature range of 20 to 35°C. He also observed that night-time temperatures below 20°C led to a delay in flowering, slowing down its cycle. High temperatures are detrimental to the growth and development of the cowpea plant, influencing flower abortion, the number and length of pods and, consequently, the number of grains per pod, thus reducing production, as well as causing phyto-illness when related to high temperatures and high air humidity (CARDOSO et al., 2005).

With regard to water requirements, caupi needs a minimum rainfall of 300 mm during its cycle in order to produce satisfactorily. Regions in which the average annual rainfall ranges from 250 mm to 500 mm are considered suitable for planting the crop without the aid of irrigation (FANCELLI E DOURADO NETO, 1997). Ellis et al. (1994) observed in a study carried out in Australia that the occurrence of rapid water deficits in the initial period of crop development can stimulate greater development of the plant's root system, however, water stress close to and before flowering can cause severe retraction of vegetative growth, limiting production. Cordeiro et *al.* (1998) found similar results with regard to water stress in the cowpea crop in a study carried out in the state of Cearà, but they also observed that the grain filling phase is severely damaged when subjected to water stress.

2.2.2 - **Soils**

The cowpea is effectively grown on practically all types of soils, especially Yellow Latosols, Red-Yellow Latosols, Red-Yellow Argissols and Fluvial Neosols. In general, the crop thrives in soils with a regular organic matter content, which are not compacted, light and deep, aerated and with medium to high fertility. However, other soil classes such as Latosols and Quartzarenic Neosols that have low fertility can also be used, when corrected with applications of chemical and/or organic fertilizers (CARDOSO, 2005).

2.2.3 - Botanical and physiological characteristics

The cowpea is an annual herbaceous plant, it is propagated by seeds, it is self-sown and its allogamy rate is considered to be very low. Its flowers are complete, with five white, yellow or violet petals, which protect the male and female organs. It has ten stamens, the end of which is a recurved stigma, moist and covered in hairs, the purpose of which is to better adhere the pollen grains. It has a narrow, elongated ovary and anthesis usually takes place in the early hours of the morning, which conditions the occurrence of pollination, which takes place predominantly by insects (TEÓFILO; MAMEDE; SOMBRA, 1999). Its root system is classified as pivoting and can reach a depth of up to 0.80 m. It is capable of symbiosis with bacteria and fungi, providing biological nitrogen fixation - BNF (MOUSINHO, 2005).

Araùjo et. *al.* (1981) stated that the photosynthetic metabolism of cowpea is classified as C3, unlike corn and sorghum crops. The plant has two types of growth habit: determinate and indeterminate, with the indeterminate genotype predominating in Brazil. The indeterminate genotype differs from the determinate genotype in that the stem does not produce an unlimited number of nodes, growing even after the inflorescence has been emitted.

There are four different classifications in terms of its cycle: super-early cycle (where maturity is reached within 60 days of sowing), early cycle (the time to reach maturity is between 61 and 70 days after sowing), medium cycle (maturity occurs between 71 and 90 days after sowing) and finally late cycle (where maturity is only reached 91 days after sowing) (FREIRE FILHO et al., 2000).

2.3 - Panicum maximum grass

According to Rocha (1991), the genus Panicum encompasses several plants, which in turn belong to the *Gramineae* family and the *Paniceae* tribe. It has an average of 81 genera and more than 1,460 species. It is possible to find this type of plant in various places around the world, as its distribution is wide across the globe, from 40° S to 50° N latitude, but its predominance is in the humid equatorial zone (20° N to 20° S), corresponding to regions such as Africa, Central and South America, northern Australia, India, southeast Asia and the Pacific Islands, at altitudes of up to 2,000 meters.

Its possible center of origin is tropical Africa, where it later spread to the subtropics of the African continent (JANK, 1995). The first specimens were introduced to Brazil at the time of slavery, via slave ships, where they served as bedding for the slaves. Once introduced into the country, it spread rapidly, giving rise to the first cultivar, Coloniao (JANK, 2003).

The *Panicum maximum* species has always had a greater prominence in Brazil, as it is highly productive and has the ability to adapt to the different regions of the country. It stood out in the agricultural sector, being responsible for a large part of cattle fattening in the country in the 1960s to 1980s and for being a forage crop with good productive results in other animal categories, such as horses and sheep, unlike other species, such as brachiaria, which are not eaten by these animals (EUCLIDES, 1999).

According to Jank (2003), studies have been carried out to improve and launch new cultivars, but the breakthrough in research came about through an agreement between the Brazilian Agricultural Research Corporation (EMBRAPA) and the Institut Français de Recherche Scientifique pour le Développment en Coopération (ORSTOM), which was signed in 1982. Following the agreement, the National Beef Cattle Research Center (CNPGC) received a collection of P. maximum consisting of 426 apomictic accessions and 417 sexual plants collected in Africa in regions that represented the natural variability of the species. Based on these studies, EMBRAPA developed new cultivars which have already been launched, Tanzania, Mombaça and BRS Zuri, which show genetic improvement in terms of increased production, resistance to pests and diseases and an increase in the number of seeds when compared to the first cultivar in the country, Coloniao (MACEDO, 2000).

Of *Panicum*'s agronomic characteristics, soil and climate requirements are the factors that impose the greatest restrictions on its productivity (SILVA, 1995). The productive range of the species is very broad, meaning that it has relatively few limitations when it comes to adapting to altitude, rainfall, temperature, etc. As far as soil fertility is concerned, the grasses of this genus are quite demanding, being suitable for sowing in soils of medium to high fertility, or after growing annual crops, followed by mineral fertilization in soils of low fertility (SANCHEZ; SALINAS, 1981).

2.3.1 - Cultivar BRS Zuri

Among the new cultivars developed by Embrapa, BRS Zuri stands out, with an average annual production of 21.8t ha^{-1} $year^{-1}$ of leaf dry matter, 50% more than 'Colonico'. Its crude protein content ranged from 11 to 15% in the leaves (SAFRASUL, 2015).

According to Embrapa Gado e Corte (2013), the cultivar is moderately adapted to soils susceptible to waterlogging, similar to what happens with the Tanzania cv. However, it develops better in well-drained soils, making it an option for diversifying pastures in biomes with more clayey soils.

It is a cespitose plant and its leaves are hairless. It has a thick stalk and internodes of medium length with little waxiness. Its sheath is moderately hairy. Its inflorescence is a large panicle with medium-long primary branches and long secondary branches only at the base. The spikelets are evenly distributed along the branches, are glabrous and have few purple spots. It has a hairy whorl at the base of the inflorescence. Its flowering is late and well defined (EMBRAPA, 2013).

2.4 - Irrigation in the semi-arid region

Agriculture depends on two essential factors of production: water and soil. The northeastern semi-arid region is characterized by intense water restriction, which imposes limits and conditions on agricultural production (BUAINAIN; GARCIA, 2015). However, water restriction does point to some positive factors, which are a lower incidence of pests and diseases and the availability of highly fertile soils with potential for irrigation (DANTAS, 2010). Irrigation, in turn, is a technology that aims to apply water to the soil, with the main objective of meeting the water requirements of the plant-atmosphere system (EMBRAPA, 2015).

The use of irrigation for agricultural production has come under serious criticism recently, mainly from those who consider it to be one of the main culprits for the problem of water scarcity in the country, especially in the semi-arid region. However, criticism of irrigation techniques based solely on the aspects mentioned above, without taking into account the importance of irrigation for food production and the Brazilian agricultural economy, can be considered unfounded or irrational (TESTEZLAF, MATSURA and CARDOSO, 2002).

In order to manage irrigation properly, some important practices must be taken into account, including the right amount of water and the right time to irrigate. By doing this, not only will the producer be saving water and energy, but he will also be ensuring better performance in terms of productivity. In addition to generating income, making rational use of water for irrigation also contributes to preserving the environment, making this technique sustainable and profitable (LAZIA, 2012).

According to Hernandez (2004), irrigation can be carried out using different methods and systems, each with its own characteristics that distinguish it from the others. The main irrigation methods are sprinkler irrigation, surface irrigation and localized irrigation, where sprinkler irrigation consists of conventional sprinkler systems, over or under the crop. In surface irrigation, the most common are flood and furrow systems and finally localized irrigation, which consists of micro-sprinkling, drip irrigation and recently drip tapes. This last method is considered to be more efficient because of its low working pressure and also because it only wets part of the soil surface, reducing water wastage.

The indiscriminate use of irrigation can cause problems for the agricultural suitability of the soil, one of the main reasons being the use of poor quality water with high saline levels, which can lead to soil salinization. This problem is very worrying as it has disastrous consequences for both the soil and crop production in general (SUASSUNA, 1996). According to Carvalho et al. (2006), in order for irrigators to avoid problems later on, they should always use good quality water and properly maintain their system, especially in the case of localized irrigation, where emitters are easily blocked by salts or deposited organic material.

Farmers find in irrigation and fertilization two elements for immediately increasing their yields, but they must always be aware of the need to choose quality seeds and seedlings, which is an essential initial factor for obtaining high yields. In irrigated agriculture, the whole process begins with the choice of the type of system to be used, so the system must be chosen carefully, looking at the competence of the company responsible for the project, its technical capacity and also its ability to provide technical assistance to the producer, as it is essential that the irrigation system purchased accompanies the farmer for a long time, in order to retain costs (HERNANDEZ, 2004).

3 - MATERIAL AND METHODS:

3.1 Location and characterization of the experimental area

The experiment was conducted in the experimental area of the Agrometeorological Station (Figure 1), belonging to the Department of Agricultural Engineering of the Federal University of Cearà (UFC), located on the Pici Campus in Fortaleza, Cearà, at the geographical coordinates 03°45'S, 38°33'W, average altitude 19 m, from September to November 2015.

Figure 1 - General view of the experimental area at the Agrometeorological Station, UFC.

Source: Author (2015).

3.2 Weather

According to the Koppen classification (1923), the study region is classified as Aw', indicating a tropical rainy climate, very hot, with a predominance of rain in the summer and fall seasons and an average temperature of over 18°C in all months. During the period of the experiment, daily observations were made of some meteorological variables, where monthly averages were obtained and are shown in Table 1.

Table 1 - Weather averages observed during the experiment.

Month	Temperature Air Average	Speed of Relative humidity	Wind	Precipitation
	(°C)	(%)	$(m\ s)^{-1}$	(mm)
September	26,7	71	4,1	6,6
October	27,7	68	4,1	1,3
November	27,4	70	4,1	1,1

Source: Agrometeorological Station, Fortaleza, Ceara, 2015.

3.3 Soil

The experiment was conducted in 20 L plastic pots. The soil used to fill the pots was Argissolo Vermelho- Amarelo (EMBRAPA, 2006), from an area of native vegetation belonging to the GEPPe - Group of Studies and Practices in Permaculture, located near the agrometeorological station.

In order to characterize the soil attributes, and with a view to better representing the soil, a composite sample was taken from the 0 - 20 cm layer and sent to the laboratory of the *Federal University of Ceará (UFC)* for physical and chemical analysis. The results obtained are shown in Tables 2 and 3.

Table 2 - Physical attributes of the red-yellow clay loam used in the experiment.

Layer	Particle size composition			Textural class	Density		Humidity (kPa)	
	Sand	Silt	Clay		Soil	Particles	33	1.500
(m)	$(g\ kg)^{-1}$			(-)	$(g\ cm)^{-3}$		$(g\ 100g)^{-1}$	
0,00 - 0,20	805	107	133	Free sand	1,49	2,69	6,43	5,33

Source: Author (2015).

Table 3 - Chemical attributes of the red-yellow clay loam used in the experiment.

Layer	pH (Agua)	CE	Assortative complex					
			Ca^{+2}	Mg^{+2}	K^{+}	Na^{+}	$H^{+} + Al^{+3}$	Al^{+3}
(m)	(-)	$(dS\ m)^{-1}$				$(cmolc\ kg)^{-1}$		
0,00 - 0,20	5,0	0,08	0,70	0,60	0,13	0,09	2,81	0,95
Layer	S	T	V	PST	M.O.	$P_{Assimilable}$		
(m)	$(cmolc\ kg)^{-1}$		(%)		$(g\ kg)^{-1}$	$(mg\ kg^{-})^{1}$		
0,00 - 0,20	1,5	4,3	35	2,0	9,83	7		

[1]pH - hydrogen potential, EC - electrical conductivity of the soil saturation extract; S - sum of bases; T - cation exchange capacity; V - base saturation; PST - percentage of exchangeable sodium; O.M. - organic matter.

Source: Author (2015).

3.4 - Water

The water used for irrigation came from a tubular well located at the agrometeorological station, with an electrical conductivity of around 0.8 dS m^{-1} , and was checked daily, pumped and stored in a reservoir with a volumetric capacity of 5000 L and pumped from there to the planting area.

3.5 - Cultures

Bean cultivar *(Vigna unguiculata)* BRS Marataoa and grass cultivar *(Panicum Maximum)* BRS Zuri were used. Both cultivars are produced by the Brazilian Agricultural Research Corporation - Embrapa.

3.5.1 - Beans BRS Marataoa

The seeds were obtained by crossing the Seridó cultivar with the TVx 1836-013J strain at

Embrapa Arroz e Feijao, located in Goiânia - GO in 1990 and 1991; its segregating generation was sent to Embrapa Meio- Norte where the strains were separated. It has a 70-75 day cycle, indeterminate growth habit, semi-prostrate size, purple dioecious flowers, greenish-white seeds and a yield of 1,200 kg ha^{-1} (Figure 2) (EMBRAPA, 2004).

Figure 2 - Illustrative images of the BRS Marataoa cultivar

Source: Embrapa Meio Norte, adapted by the author.

The bean crop cycle in this research was 75 days, grown in full sun, and harvested by hand as the pods ripened.

3.5.2 - BRSZuri grass

This cultivar is the result of a massai selection on *Panicum maximum* derived populations from Tanzania, East Africa. The selection work was coordinated by Embrapa Gado de Corte in partnership with other institutions. The selection parameters were based on higher productivity, vigor and carrying capacity. It is a cespitose plant, erect and tall, with long, dark green, arched and wide leaves. It has high nutritional value, high yields and resistance to leaf spot caused by the fungus *Bipolaris maydis* and pasture leafhoppers. Figure 3 shows some of these characteristics. The cycle of the BRS Zuri grass in this research was 70 days, the beginning of flower ripening, but due to the different times when they were transplanted into pots, they were harvested together with the end of the bean cycle, 75 days after the start of the experiment.

Figure 3 - Illustrative images of Cultivar BRS Zuri, seeds and clumps.

Source: Embrapa Gado de Corte, adapted by the author.

3.6 - Experimental design and treatments

A split-plot randomized block design was used, with three replications. The plots were made up of 5 irrigation levels, equivalent to 0.25, 0.50, 0.75, 1.0 and 1.25 times the water evaporation measured in the Class A tank (ECA) and the sub-plots were made up of when the grass was transplanted (at the time of bean emergence, EF; 5 days after bean emergence, DAEF; 10 DAEF; and 20 DAEF), (Figure 4).

Figure 4 - Sketch showing the distribution of treatments in the experimental area.

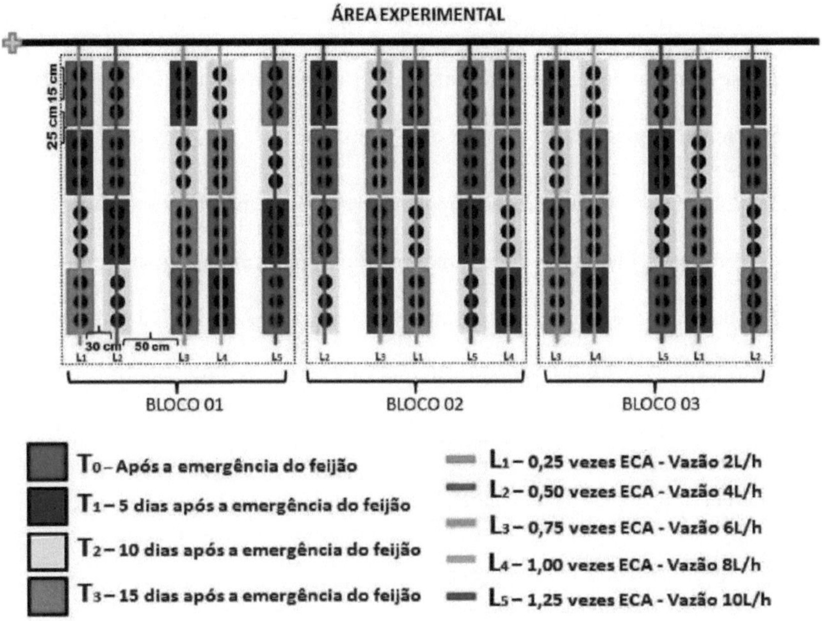

Source: Prepared by the author (2015).

Each subplot contained three pots, each cultivated with the two plants in the consortium (beans and grass), making up 60 experimental units per block, totaling 180 pots in the experimental area.

1.7 - Preparing the experimental area

The area used to conduct the experiment was 9.6 x 6.9 m, i.e. 66.24 m^2 . Initially, the 180 flexible plastic pots were distributed according to the experimental design (the pots had circular holes in the bottom and were placed on bricks to allow excess water to drain away). They were then filled with the soil described above to plant the consortium according to the proposed treatments.

3.7.1 *Setting up the irrigation system*

A localized drip irrigation system was used, consisting of a main line made up of a PVC pipe with a nominal diameter of 50 mm and a length of 12 m; with 15 lateral lines made up of polyethylene hoses with a diameter of 25 mm, one for each row of pots. The system was pressurized by a motor pump set with a power of 0.75 hp.

Once the lateral lines had been laid out, the drippers were inserted into them according to the flow rate needed for each line to have the same irrigation time, depending on the different treatments. For this purpose, Netafin (2, 4 and 8 L h^{-1}) and Agrojet (10 L h^{-1}) self-compensating drippers were used (Figure 5). In order to achieve a flow rate of 6 L.h^{-1} , a combination of a 2 and 4 L h^{-1} dripper was used.

Figure 5 - Different drippers used on the lateral lines.

Source: Netafin; Agrojet. Adapted by the author (2015)

Next, the system's uniformity test was carried out using the Keller; Karmeli (1974) method, which recommends obtaining the flow rates at four points along the lateral line, the first, 1/3, 2/3 and last dripper; however, the flow rates of all the drippers were collected, adapting the methodology to the needs of the experiment, since the drippers had different flow rates.

The flow rate was obtained from the ratio between volume and time. The volume was collected using collectors over a period of 5 minutes and measured using 500 mL graduated beakers; the time was observed using digital chronometers. For greater reliability, the process was repeated three times for each dripper and averaged to obtain the final value.

After measuring the flow in all the drippers, the system was evaluated, following an adaptation of the Keller; Karmeli (1974) methodology. CUC (equation 1), CUD (equation 2), CUE (equation 3) and Ef (equation 4) were calculated for each flow rate. To evaluate the system as a whole, the average values of all the flow rates were obtained and according to Mantovani's classification (2001), the system was classified as excellent, with approximately 90% efficiency.

$$CUC = \left(1 - \frac{\sum \lfloor X_i - \overline{X} \rfloor}{n.\overline{X}}\right) \cdot 100 \qquad (1)$$

in which:

- CUC - Christiansen's coefficient of uniformity (%);
- X_i - flow rate of each dripper, L h ;$^{-1}$
- $\bar{X}$ - average dripper flow rate, L h ;$^{-1}$
- n - number of drippers

$$CUD = 100 \frac{q_{25\%}}{\overline{q}} \qquad (2)$$

in which:

- CUD - coefficient of distribution uniformity, (%);
- $q_{25\%}$ - average flow rate of the 25% lowest qi values, L h ;$^{-1}$
- q - average of all drippers, L h^{-1}

$$CUE = 100\left(1 - \frac{S_q}{\overline{q}}\right) \qquad (3)$$

in which:

- *CUE* - statistical uniformity coefficient, (%);
- *sq* - standard deviation of flows at emission points.
- *q* - average of all drippers, L h^{-1}

$$E_f = Tr \cdot CUC \quad (4)$$

in which:

- E_f- Efficiency of the drip irrigation system
- Tr - Transmissivity coefficient according to soil texture, climate and root depth. Constant = 0.95;
- CUC - Christiansen uniformity coefficient (%);

3.7.2 Fertilization

Fertilization was based on the results of the soil analysis and the nutritional recommendations for *Vigna unguiculata* bean crops and *Panicum maximum* grass, as prescribed by Aquino *et* al. (1993). Initially, mineral fertilization was applied directly to the pots one week before sowing the bean seeds, quantified according to the following information.

The amounts of macronutrients, according to the recommendation, were 20 kg ha^{-1} of Nitrogen (N) for both crops, 80 and 60 kg ha^{-1} of Phosphorus (P), 20 and 30 kg ha^{-1} of Potassium (K) for beans and grass, respectively.

We proceeded as suggested by Braga (2013): when intercropping in pots, you should opt for the highest recommendation in order to maintain a balance between the two intercropped crops. As a result, 20 kg ha^{-1} of N, 80 kg ha^{-1} of P and 30 kg ha^{-1} of K were used.

Because the research was carried out in pots, it was necessary to adapt the fertilizer recommendation from an area basis to a volume basis. To do this, the volume of the pots was calculated according to Equation 5

Figure 6 - Sizing the volume of the pot for fertilization

Source: Braga, adapted by the author (2013)

$$Volume = \pi . h . \left(\frac{R1^2 + R1.R2 + R2^2}{3} \right) \quad (5)$$

Once we knew that the volume of the pot was 26.5 dm^3 , we converted the area into volume. Knowing that 1 ha is equivalent to 10,000 m2 and the distribution layer of the mineral fertilizer is 20 cm, in this case 1 ha is equivalent to 2,000,000 dm^3 . Once the volume was known, the quantities of nutrients to be applied per pot were calculated, as shown in Table 4.

Table 4 - Fertilizer recommendations and amount of fertilizer used

	Fertilizer recommendations		
	Recommendation ha^{-1}	Recommendation per pot (mg)	Amount of fertilizer used
Urea (N)	20 kg ha^{-1}	265 mg pot^{-1}	350 mg pot^{-1}
Super Phosphate Simple (P2O5)	80 kg ha $^{-1}$	1,060 mg pot^{-1}	6,240 mg pot^{-1}
Potassium Nitrate	30 kg ha $^{-1}$	397.5 mg vessel^{-1}	903 mg pot $^{-1}$

(K2O)

Source: Prepared by the author (2016).

To weigh the nutrients individually, a digital scale with an accuracy of 0.005 g was used, as shown in Figure 7. The sources of N, P and K used were: Urea, Super Simple Phosphate and Potassium Nitrate, respectively.

After weighing, the fertilizers were incorporated into the soil along with the substrate in a 4:1 ratio (four parts soil to one part tanned cattle manure) and turned over in a 0.20 m layer. Then 500 mL of water was added manually to each pot, measured in a graduated beaker to dissolve them.

Figure 7 - Weighing fertilizer on a precision scale.

Source: Author (2015)

3.8 - Sowing and thinning

Depending on the treatment, the grass was transplanted at different times in relation to the emergence of the beans. To do this, the grass was planted in stages, every 5 days, in four 162-cell expanded polyethylene trays (Figure 8). Transplanting for each season, in accordance with the treatments, was carried out when the grass plants had been sown for 20 days, corresponding on average to 13 days after grass emergence; it is worth noting that the grasses were transplanted at the same age.

Figure 8 - BRS Zuri grass seedlings in expanded polyethylene trays at 20, 15, 10 and 0 days after sowing.

Source: Author (2015).

The beans were sown directly into the pots, three seeds per hole, at a depth of 5 cm. After 5 days, 100% of the bean seeds had emerged and the first BRS Zuri grass seedlings were transplanted into the pots. At 10 days after the beans emerged, the second grass seedlings were transplanted and the beans were thinned out, leaving just one bean plant per pot.

The other BRS Zuri seedlings were transplanted 5 days apart, except for the seedlings in the last tray, which were transplanted 20 days after the beans emerged.

3.9 - Irrigation management

Irrigation was carried out daily, using climate management. The methodology proposed by Bernardo, Soares and Mantovani (2006) was used to estimate the crop's potential evapotranspiration (ETpc), where the reference evapotranspiration (ETo) was first estimated.

The daily estimate of reference evapotranspiration (ETo) was made using a Class A evaporimetric tank belonging to the Agrometeorological Station of the Federal University of Ceará, located next to the experimental area, with the water evaporation in the tank quantified daily at 9 am.

To estimate the reference evapotranspiration - ETo, equation 6 was used

$$ETo = ECA.Kt \quad (6)$$

in which:

ETo - reference evapotranspiration estimated using the Class A Tank (mm);

ECA - evaporation measured in the Class "A" tank (mm);

Kt - tank adjustment coefficient (dimensionless factor).

The tank adjustment coefficient (Kt) was obtained using equation 7 proposed by SNYDER (1992):

$$Kt = 0{,}482 + 0{,}024\ln(F) - 0{,}000376\,U + 0{,}0045\,UR \quad (7)$$

in which:

F - distance from the center of the tank to the edge of the border (tank on grass cover 10 cm from the ground);

U - average wind speed at 2 m height (km day);$^{-1}$

RH - average relative humidity (%).

To obtain (Kt), we used the arithmetic averages of the weather data recorded by the UFC's agrometeorological station for the months during which the experiment was conducted over the last five years.

Equation 8 was used to estimate the potential evapotranspiration of the crop (ETpc).

$$ETpc = ETo\ .Kc \tag{8}$$

in which,

ETpc - potential crop evapotranspiration (mm);

ETo - reference evapotranspiration estimated using the Class A Tank (mm);

Kc - crop coefficients;

The crop coefficient was adopted according to the phenological stages of the bean crop, since it was present in all the pots from the beginning of the treatments. The coefficients used were 0.70 (up to 12 days after sowing, DAS); 0.81 (from 13 to 33 DAS); 1.2 (from 34 to 54 DAS) and 0.77 (from 55 DAS to the end of the cycle), according to Sousa, Bezerra and Teófilo (2005).

Equation 9 was used to calculate the irrigation time.

$$\mathrm{Ti} = \frac{\mathrm{ETpc}\ .\ A_v\ .\ F_c}{E_i\ .\ f_i} \tag{9}$$

Where,

Ti - is the irrigation time (h);

f_i - Dripper flow rate (L/h)

ETpc - potential crop evapotranspiration (mm);

Av - vessel area (0.00313 m);[2]

Fc - soil cover factor (dimensionless);

Ei - irrigation efficiency (90%) (dimensionless);

3.10 - Phytosanitary control and weeds

Weeds present during the experiment were controlled weekly by hand weeding. This avoided competition for water and nutrients with the intercropped crops.

During the experiment, the presence of fleas *(Aphis craccivora)* was constant, *and* these were controlled by spraying every 15 days with the insecticide Malathion®, registered with the Ministry of Agriculture, Livestock and Supply, for growing beans (AGROFIT 2008). The product was used at a concentration of 2.5 mL L^{-1} of water and applied via spraying.

3.11 - Variable analyzed

At 40 days after emergence, the relative chlorophyll index was determined using a SPAD-502 chlorophyll meter on the second fully expanded leaf from the apex to the base of the plant in order to assess the effects of the different treatments on the plant of interest, in this case the cowpea.

3.12 - Statistical analysis

The results obtained were subjected to analysis of variance, with the treatments with quantitative variables being subjected to regression analysis and the treatments with qualitative variables being subjected to the Tukey test at 1 and 5% significance levels.

4 - RESULTS AND DISCUSSION:

From the results shown in Table 5, there was a significant influence at the 1% probability level by the F test of the irrigation rates (L) and grass insertion times (E) on the variable analyzed, with an interaction between the treatments at the 1% level.

Table 5 - Summary of the analysis of variance for the chlorophyll content present in cowpea plants subjected to different irrigation rates (L) and grass transplanting times (E), Fortaleza, Ceará 2016.

Variation source	GL	SQ	QM	F
Blocks	2	39.22234	19.61117	2.1475 ns
Blades of Irrigation (L)	4	4613.94958	1153.48740	126.3087 **
Residue-(L)	8	73.05828	9.13229	
Installments	14	4726.23021	-	-
Insertion time (E)	3	23235.42050	7745.14017	1042.0802 **
Int. W x H	2	1364.55416	113.71285	15.2996 **
Residue-(E)	30	222.97152	7.43238	
CV % (Làmina)		3.74		
CV % (Season)		3.37		
Total	59	29549.17640		

Source: Author (2016)

When analyzing the effect of the interaction between irrigation rates and grass transplanting times on the variable studied, it was observed that the mathematical model that best represented the data for this variable was the polynomial type. The coefficients of determination obtained were (R^2) of 0.9656 for insertion time 0 (at the time of bean emergence), (R2) of 0.7178 for insertion time 5, (R?) of 0.9788 for insertion time 10 and (R^2) of 0.9333 for insertion time 20 days after bean emergence, DAEF (Figure 9).

Figure 9 - Chlorophyll content in beans at 40 DAE when subjected to irrigation rates at different grass transplanting times. Fortaleza, Ceará, 2016.

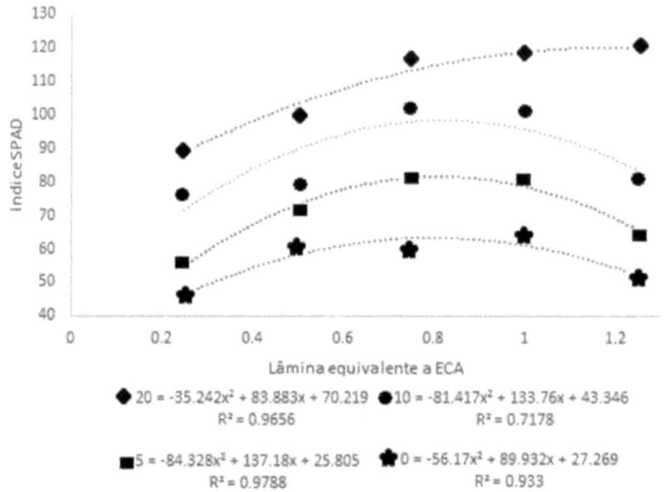

Source: Authors (2016).

There was an increase in chlorophyll production in the Caupi bean as water availability increased with the irrigation provided at all transplanting times. It is also possible to see in Figure 9 that in relation to the grass insertion periods, there was a more pronounced growth in the line representing the 20 DAEF treatment. Looking at the treatments in which the grass was transplanted at emergence and 5 DAEF, it can be seen that, compared to the other treatments, they produced the lowest chlorophyll levels, which can be explained by the fact that the bean plant competes with the grass plant for nutrients for a longer period of time. Therefore, it can be said that the chlorophyll content of the beans increased when the irrigation water was increased and when the time of grass insertion was delayed.

The higher values found in the treatment in which the grass was transplanted at 10 and 20 DAEF compared to the other treatments can be explained by the fact that the longer transplanting time allowed for better nutritional and water conditions for the bean crop, due to less competition in the cycle with the grass, as well as greater availability of sunlight for the plants in this treatment. In a study carried out by Araùjo et *al.* (1986), it was observed that there was also a decrease in phytomass production in the bean crop, and consequently the plants had lower chlorophyll indices when subjected to intercropping with grasses, due to their rapid development. Costa and Silva (2006) explain this decrease as a result of competition between the crops for light, nutrients and water.

5 - CONCLUSIONS:

Observing the effect of the blades evaluated, it can be concluded that the blades equivalent to 0.75 and 1.00 times ECA showed the highest chlorophyll index in the leaves of the Caupi bean. As for the treatment of grass insertion time, the 10 and 20 DAEF times stood out, showing that the legume's competition with the grass directly affects chlorophyll production in the bean plant.

6 - REFERENCES

AGROFIT (Sistema de Agrotóxicos Fitossanitários) - **Commercial products registered for aphid control on bean crops,** January 2016. Available at <http://agrofit.agricultura.gov.br/agrofit cons/!ap produtoolorm detalhe cons?p id produto Iormulado tecnico 5469&p tipo janela=NEW>. Accessed on 05 Jan 2016.

AIDAR, H.; VIEIRA, C. Associated cultivation of beans and corn. III. Effect of plant populations in the simultaneous planting system of both crops. **Revista Ceres**, v.26, n.143, p.102-111, 1979.

ALVARENGA, R. C., NOCE, M. A.. **Crop-livestock integration**. Sete Lagoas: Embrapa Milho e Sorgo, 2005.

ANDRADE JÛNIOR, A. S.; RODRIGUES, B. H. N.; BASTOS, E. A. Irrigation. In: CARDOSO, M. J. (Org.) **A cultura do feijâo caupi no Meio-Norte do Brasil.** Teresina: Embrapa Meio-Norte, 2000. p. 127-154. (Embrapa Meio-Norte. Technical Circular, 28).

ANDRADE JÛNIOR, A.S. de; SANTOS, A.A. dos; ATHAYDE SOBRINHO, C.; BASTOS, E.A.; MELO, F. de B.; VIANA, F.M.P.; FREIRE FILHO, F.R.; CARNEIRO, J. da S.; ROCHAS, M. de M.; CARDOSO, M.J.; SILVA, P.H.S. da; RIBEIRO, V.Q. **Cultivation of cowpeas** (Vigna uguiculata (L.) Walp.). Teresina, PI: Embrapa Meio-Norte, 2002. 108. (Embrapa Meio-Norte. Sistema de Produçao: 2)

ANDRADE, C.M.S.; CARNEIRO, J.C.; VALENTIM, J.F. et al. Effect of shading on dry matter accumulation rates of four forage grasses. In: ANNUAL MEETING OF THE BRAZILIAN SOCIETY OF ZOOTECHNICS, 39, 2002, Recife. **Anais**... Recife: Brazilian Society of Animal Science, 2002.

ANDRADE, C.M.S.; GARCIA, R.; COUTO, L. et al. Factors limiting the growth of tanzania grass in an agroforestry system with eucalyptus, in the Cerrados region of Minas Gerais. **Revista Brasileira de Zootecnia**, v.30, n.4, p.1178-1185, 2001 (supplement).

ANDRADE, CMS de et al. Performance of six single grasses or grasses intercropped with Stylosanthes guianensis cv. Mineirao and eucalyptus in a silvipastoral system.

Revista Brasileira de Zootecnia, v. 32, n. 6, p. 1845-1850, 2003.

ANDRADE, M. A. de.; RAMALHO, M. A. P.; ANDRADE, M. J. B. de. Intercropping of beans (Phaseolus vulgaris L.) with corn (Zea mays L.) cultivars of different sizes. **Agros**, v.4, n.2, p.23-30, 1974.

AQUINO, A. B., AQUINO, B. F., HERNANDEZ, F. F. F., HOLANDA, F. J. M., FREIRE, J. M., CRISOSTOMO, L. A., ... & FERNANDEZ, V. Recomendações de adubaçao e calagem para o estado do Cearâ. **Fortaleza: UFC**, 1993, 248p.

ARAÙJO, G., FONTES, L., & LOPES, N. Production and production components in associated and exclusive corn and bean cropping systems. **Revista Ceres, Viçosa**, v. 33, n. 190, p. 469-478, 1986.

ASSMANN, A.L. et.al., Beef cattle production and dry matter accumulation in a crop-livestock integration system in the presence and absence of white clover and nitrogen. **Revista Brasileira de Zootecnia**, v. 33, n. 1, p. 37-44, 2004.

BARBOSA, R. A., NASCIMENTO JÛNIOR, D. D., EUCLIDES, V. P. B., SILVA, S. D., ZIMMER, A. H., & TORRES JÛNIOR, R. D. A.. Tanzania grass subjected to combinations of grazing intensity and frequency. **Pesquisa Agropecuâria Brasileira**, v. 42, n. 3, p. 329-340, 2007.

BASTOS, E.A. **Adaptation of the cropgro model for simulating the growth and development of cowpea (Vigna unguiculata (L.) Walp.) under different water conditions in the state of Piaui**. 1999. Doctoral thesis.

BERNARDO, S.; SOARES, A. A.; MANTOVANI, E. C. **Manual de Irrigaçâo**. 8ª Ed. Viçosa, Ed. UFV, 2006, 625p.

BEZERRA, A. P. A. et. al., Yield, production components and efficient land use in sorghum x cowpea and sorghum x maize consortia. **Revista Ciência Agronômica**, v. 38, n. 1, p. 104-108, 2008.

BINKLEY, D.; GIARDINA, C. Nitrogen fixation in tropical forest plantations. In: NAMBIAR, E.K.S.; BROWN, A.G. (Eds.) **Management of soil, nutrients and water in tropical plantation forests.** Australia: ACIAR; CSIRO; CIFOR,1997. p.297-337.

BUAINAIN, A. M., & GARCIA, J. R. Polos de Irrigação no Nordeste do Brasil. desenvolvimento recente e perspectivas. **Confins. Revue franco-brésilienne de géographie/Revista franco-brasilera de geografia**, n. 23, 2015.

BUXTON, D.R.; FALES, S.L. Plant environment and quality. In:FAHEYJUNIOR, G.C. (Ed.). Forage quality, evaluation and utilization. Lincoln: University of Nebraska, 1994. p.155-199.

CARDOSO, M. J.; MELO, F. de B.; LIMA, M. G. de. **Ecophysiology and planting management.** In: FREIRE FILHO, F. R.; LIMA, J. A. de A.; RIBEIRO, V. Q. (Ed.). Cowpea: technological advances. Brasilia: Embrapa Informaçao Tecnologia, 2005. p.212-228.

CARDOSO, M. J. et al. BR 12-Canindé": early macassar bean cultivar with multiple virus resistance. **Teresina: EMBRAPA-UEPAE de Teresina**, 1988.

CARVALHO, de Carvalho, C. M., Eloi, W. M., Lima, S. C. R. V., & Pereira, J. M. G. Performance of a drip irrigation system in guava cultivation. **Irriga**, v. 11, n. 1, 2006.

CORDEIRO, L. G., BEZERRA, F. M. L., SANTOS, J. J. A., & MIRANDA, E. P.. Sensitivity factor to water deficit in the Caupi bean crop (Vigna unguiculata (L). Walp.). **Revista Brasileira de Engenharia Agricola e Ambiental, Campina Grande**, v. 2, n. 2, p. 153-157, 1998.

COSTA, A. S. V., & SILVA, M. B. Corn-bean consortium systems for the rio doce valley region, Minas Gerais. **Ciênc. agrotec.**, v. 32, n. 2, 2008.

COSTA, D. S., & BARBOSA, R. M. PRODUCTION SYSTEMS AND BEAN CULTIVARS IN INTERCROPPING WITH MAIZE1 PRODUCTION SYSTEMS AND BEAN CULTIVARS IN INTERCROPPING WITH MAIZE.

Scientia Agraria, v. 11, n. 6, p. 425-430, 2010.

CROOKSTON, R. K.; TREHARNE, K. S.; LUDFORD, P.;OZBUN, J. L. Response of beans to shading. **Crop Science**, v.15, p.412-416, 1975.

DA ROSA, S. R. A., DE AQUINO PORTES, T., & DE OLIVEIRA, I. P.. Growth analysis of Tanzania grass in single planting and legume consortium systems. **Ciência Animal Brasileira**, v. 8, n. 2, p. 251-260, 2007

DANTAS, E. W. C. Changes in the Brazilian Northeast: a reflection on food production and hunger in contemporary times. **Confins. Revue franco-brésilienne de géographie/Revue franco-brasilera de geografia**, n. 10, 2010

DETOMINI, E. R.; MONTEIRO, R. O. C.; COELHO, R. D. Evaluation of the production of

Panicum maximum cv. Tanzania under different levels of environmental supply.

Pasturas Tropicales, v. 27, p. 26-37, 2005. TAIZ, L.; ZEIGER, E. **Fisiologia vegetal**. Porto Alegre: Artmed, 2004. p. 449-484.

DIECKOWIV, A.P.J. Crop-livestock integration: intensification of agricultural land use. **Ciência Rural**, v. 39, n. 6, 2θθ9.

EMBRAPA - **BRS Zuri, Production and resistance for livestock,** April 2014.

Available at <

http://ainfo.cnptia.embrapa.br/digital/bitstream/item/123642/1/Folder-Zuri-Final- 2014.pdf >. Accessed on: Jan. 5, 2015.

EMBRAPA - **BRS Zuri, production and resistance for livestock**, April 2014. Available at <https://www.embrapa.br/documents/1355008/1528459/Folder+Zuri.pdf/e89a784d- fe75-47ff-8a79-6065f85b8fb5>. Accessed on 10 Jan 2016.

EMBRAPA, National Soil Research Center (Rio de Janeiro, RJ). **Brazilian Soil Classification System**.3ª Ed. Rio de Janeiro, 2013, Ed. Embrapa 353p.

EMBRAPA. - **Soil and water management**, January 2016. Available at < http://www.cpatsa.embrapa.br:8080/pesquisa/manejo.html >. Accessed on January 15, 2016.

EUCLIDES, V. P. B., THIAGO, L. R. L. D. S., MACEDO, M. C. M., & OLIVEIRA, M. P. D. Voluntary intake of three cultivars of Panicum maximum under grazing. **Revista Brasileira de Zootecnia**, v. 28, n. 6, p. 1177-1185, 1999.

FANCELLI, A. L., & Neto, D. D. (Ed.). **Irrigated bean production technology**. Fealq, 1997.

FOLEGATTI, M. V., PAZ, V., & OLIVEIRA, A. Yield of irrigated bean submitted to different levels of water with furrow irrigation. **Revista Brasileira de Engenharia Agricola e Ambiental**, v. 3, n. 3, p. 281-285, 1999.

FREIRE FILHO, F. R., CARDOSO, M. J., & DE ARAÙJO, A. G. Caupi: scientific nomenclature and common names. **Pesquisa Agropecuâria Brasileira**, v. 18, n. 12, p. 1369-1372, 1983.

FREIRE FILHO, F. R. Genética do caupi. **The cowpea in Brazil**, 1988.

FREIRE FILHO, F. R.; LIMA, JA de A.; RIBEIRO, V. Q. **Feijâo-caupi: technological advances**. Embrapa Informaçâo Tecnológica; Teresina, PI: Embrapa Meio-Norte, 2005.

FREIRE FILHO, F. R.; RIBEIRO, V. Q.; BARRETO, P. D.; SANTOS, A. A. dos. Cowpea cultivars for the Mid-North region of Brazil. **Cowpea cultivation in the mid-north of Brazil. Teresina: EMBRAPA Meio-Norte**, p. 67-88, 2000.

HERNANDEZ, F. B. T. **Irrigation management**. 2004. Available at <http://www.irrigaterra.com.br/manejo.php>. Accessed on January 15, 2016.

IBGE, **Systematic survey of agricultural production**. Rio de Janeiro. v. 5, n. 12, 1993; v. 6, n. 12, 1994; v. 7, n. 12, 1995; v. 8, n. 12,1996; v. 9, n. 12, 1997; v. 10, n. 12, 1998; v. 11, n. 12, 1999; v. 12, n.12, 2000; v. 13, no. 12, 2001; v. 14, no. 12, 2002; v. 15, no. 12, 2003; v. 16, no. 12, 2004; v. 17, no. 12, 2005; v. 18, no. 12, 2006; v. 19, no. 12, 2007; v. 20, n. 12, 2008; v. 21, n. 12, 2009; v. 22, n. 12, 2010; v. 23, n. 12, 2011; v. 24, n. 12, 2012; v. 25, n. 12, 2013; v. 26, n 12, 2014.

JANK, L. The history of Panicum maximum in Brazil. **Revista JC Maschietto, Campo Grande**, v. 1, n. 1, p. 14, 2003.

JANK, L. Improvement and selection of Panicum maximum varieties. **Symposium on pasture management**, v. 12, p. 21-58, 1995.

JUNIOR, A. B., MORAES, A., VEIGA, M., PELISSARI, A., & DIECKOW, J.

Crop-livestock integration: intensifying the use of agricultural land. **Cienc. RuralV 39**, p. 1925-1933, 2009.

LAZIA, B. - **The importance of irrigation for productivity**, May 2012.

Available at: < http://www.portalagropecuario.com.br/agricultura/irrigacao/a- importance-of-irrigation-for-productivity/ >. Accessed on 15 Jan 2016.

LIMA, L. A. de P.; VIEIRA, C. **Associated cultivation of corn and beans.** IV. Behavior of production systems. In: Bean Project: Report 78/79. Belo Horizonte, p.27-32, 1982.

LOURENÇO, L. F. Evaluation of tanzania grass production in a protected environment under variable availability of water and nitrogen in the soil. **Evaluation of tanzania grass production in a protected environment under variable availability of water and nitrogen in the soil**, 2004.

MACEDO, M. C. M., EUCLIDES, V. P. B., VALÉRIO, J. R., & BONO, J. A. M. Cultivar Massai (Panicum maximum) a new forage option: adaptation and productivity characteristics. **ANNUAL MEETING OF THE BRAZILIAN SOCIETY OF ZOOTECHNICS**, v. 37, p. 2000, 2000.

MACHADO, C. M. N.; FLECK, N. G.; SOUZA, R. S. Eficiência na utilização da terra rendimento das culturas em consórcio. **Pesquisa Agropecuâria Brasileira**, v.19, n.3, p.317-327, 1984.

MAIA, F. M. M. **Composition and nutritional characterization of three cultivars of Vigna unguiculata (L.) Walp: EPACE-10, Olho de ovelha and IPA-206**.

Fortaleza, 1996. 87 p. Dissertation (Master's Degree in Plant Biochemistry), Federal University of Ceará.

McKENZIE, B.A.; HAMPTON, J.G.; WHITE, J.G.H. et al. Annual crop production principles. In: WHITE, J.; HODGSON, J. (Eds.) **New Zealand pasture and crop science** . Oxford: Oxford University Press, 1999. p.199-212.

MELO, F. de B.; CARDOSO, M.J.; SALVIANO, A.A.C. Soil fertility and fertilization. In: FREIRE FILHO, F.R.; LIMA, J.A. de A.; RIBEIRO, V.Q. (Ed.). **Cowpea: technological advances**. Brasilia: Embrapa Informaçao Tecnològica, 2005. p.229-242.

MOHAMED SALLEN, M.A.; FISHER, M.J. Role of ley farming in crop rotations in the tropics. In: INTERNATIONAL GRASSLAND CONGRESS, 17., 1993, **RockhamptonTroceedings ...** Rockhampton: 1993. p.2179-2187.

MOUSINHO, F. **Economic viability of cowpea irrigation in the state of Piaui. 2005**. 2005. Doctoral thesis. Thesis (Doctorate)-Escola Superior de Agricultura "Luiz de Queiroz", Universidade de Sao Paulo, Piracicaba.

MYERS, R.J.K.; ROBBINS, G.B. Sustaining productive pastures in the tropics. 5. Maintaining productive sown grass pastures.**Tropical Grasslands**, v.25, p.104-110, 1991.

PORTE, T. A.; CARVALHO, J. R. P. Leaf area, solar radiation, air temperature and yields in intercropping and monocropping of different maize and bean cultivars. **Pesquisa Agropecuària Brasileira**, v.18, n.7, p.62-755, 1983.

QUIN, F. M. Introduction. In: SINGH, B. B.; MOHAN RAJ, D. R.; DASHIELL, K. E.; JACKAI, L. E. N. (Ed.). **Advances in cowpea research.** Ibadan: IITA; Tsukuba: JIRCAS, 1997. p. ix-xv.

RAPOSO, J.A. A. et al. Corn and bean intercropping in different arrangements and plant populations in Pelotas, RS. **Area de Informaçâo da Sede-Article in an indexed journal (ALICE)**, 1997.

RESENDE, G.O., TÀVORA, F.J.A.F.T, PAULA, P.H.F., Effect of exclusive and consortium planting systems on weed incidence and control. **Revista ciência Agronômica**, v.02, n. 1, p. 23-33, 1897

RIBEIRO, V.Q. **Cultivation of cowpeas (Vigna unguiculata (L.) Walp)**. Production Systems 2. EMBRAPA, 2002. Available at: < http://www.cpamn.embrapa.br/publicacoes/new/sistemaproducao/sistemaproducao pdf/sistemaproducao 2.pdf > Accessed on January 14, 2016.

ROCHA, G.D. Ecossistemas de pastagens: aspectos dinâmicos. **Piracicaba: FEALQ**, 1991. 391 p.

SAFRASUL SEMENTES - **Panicum maximum, cv. BRS Zuri**, January 2016. Available at <http://www.safrasulsementes.com.br/?conteudo=biz catalogo&canal id=3&pro id =26>. Accessed on 08 Jan 2016.

SANCHEZ, P.A.; SALINAS, J.G. Low input technology for managing oxisols and ultisols in Tropical America. **Advances in Agronomy**, v. 34, p. 279-405, 1981.

SILVA, S.C. Edapho-climatic conditions for the production of Panicum sp. In: SIMPÓSIO

SOBRE MANEJO DA PASTAGEM, 12, Piracicaba, 1995. **Proceedings**. Piracicaba: FEALQ, 1995. p. 129-146.

SINGH, B. B.; EHLERS, J. D.; SHARMA, B.; FREIREFILHO, F. R. Recent progress in cowpea breeding. In: FATOKUN, C. A.; TARAWALI, S. A.; SINGH, B. B.; KORMAWA, P. M.; TAMO, M. (Eds) **Challenges and Opportunities for Enhancing Sustainable Cowpea Production. International Institute of Tropical Agriculture**, Ibadan, Nigeria, p.22-40, 2002.

STONE, L. F., & MOREIRA, J. A. A. Bean crop response to nitrogen in top dressing, under different irrigation rates and soil preparations. **Pesquisa Agropecuâria Brasileira**, v. 36, n. 3, p. 473-481, 2001

SUASSUNA, J. **The salinity of waters in the semi-arid northeast.** 1996. Available at <http://www.fundaj.gov.br/index.php?option=com_content&view=article&id=770&Itemid=37>. Accessed on January 15, 2016.

TÉOFILO, E. M.; MAMEDE, F. B.; SOMBRA, N. S. Natural hybridization in caupi (Vigna unguiculata (L.) Walp-Fabaceae). **Ciência e Agrotecnologia, Lavras**, v. 23, p. 1010-1011, 1999.

TESTEZLAF, R. MATSURA, E.E & CARDOSO, J.L. **Importância da irrigaçâo no desenvolvimento do agronegócio**, July 2002. Available at < http://www.agr.feis.unesp.br/csei.pdf > Accessed on 10 Jan 2016.

FEDERAL UNIVERSITY OF CAPINA GRANDE. Assistat - Assistência Estatistica, V - 7.7 beta. Campina grande, PB, 2011. Available for download at

<http://www.assistat.com/indformp.html.> Accessed on: 12. Dec. 2015.

WILLEY, R. W.; OSIRU, D. S. O. Studies on mixtures of maize and beans (Phaseolus vulgaris) with particular reference to plant population. **Journal of Agricultural Science**, v.79, p.29-571, 1972.

FSC
www.fsc.org
MIX
Papier aus verantwortungsvollen Quellen
Paper from responsible sources
FSC® C105338

Printed by Books on Demand GmbH, Norderstedt / Germany